Progressive Pathway Volume 1

Visit www.booksurge.com to order additional copies.

CORBETT M. KROEHLER

PROGRESSIVE PATHWAY VOLUME 1
THE BEGINNING OF THE END 2008

2007

Progressive Pathway Volume 1

TABLE OF CONTENTS

Introduction   xiii
Part 1—*In Knots At 37,000 Feet*   1
   Melancholy Rambling   2
   Pain-Free Dentistry   4
   The Happiest Place On Earth   5
   The Deluge   6
   Lighter Than Air   7
   Absence Of Bluster   8
   Only Wonks Need Apply   10
   Start With The Base And Work Outward   17
   Face Time   18
   Blue Dog In A Red State   20

Part 2—*Toxins And Transplants*   21
   Surrender In Paradise   21
   Apathy, Apathy, Apathy   26
   Apathy, Global Warming, Apathy, Global
   Warming, Apathy, Global Warming   27
   Slow Death Or Deep Change   29

Part 3—*Nonnegotiables*   33
   Vapid And Breathless   33
   Stand For Something Or Fall For Anything   37
   Global Climate Change   38
   Poverty   39
   Runaway Executive Compensation   42

Reproductive Freedom   44
Love Decriminalized   45
A Politicized Pentagon   47
Public Education   50
Iraq   52
Habeus Corpus   54
Torture   56
Global Warming   57
Political Action Committees   60
$100 Per Month   63

Part 4—*Conclusion*   65

Dedicated To All Americans Who Serve Their Country, Especially Those Who Have Died Or Been Wounded In The Performance Of Their Duty.

INTRODUCTION

"What is government itself but the greatest of all reflections of human nature? If men were angels, no government would be necessary."

—James Madison

Imagine a world in which the American colonies remain just that, British possessions. We might well use the word *lift* in place of *elevator* and *standing in queue* rather than *waiting in line.* Anglicanism would be the official state religion and George Washington barely would be remembered. Thankfully for the world's sake, events took a turn for the better.

Why is the United States of America the most enduring democracy in history? If you believe in the concept of Manifest Destiny, than you already have your answer. On the other hand, if you affix no specific principle to our lasting success, where do you pin our good fortune?

America is America and will remain the land of the free and the home of the brave for generations to come because of the brilliance, courage and wisdom of our Founding Fathers. It is just that simple. They examined the history of constituent-based government as far back as Roman times and contrasted it against the tyranny under which we had suffered from the British crown. Their research and hypotheses led them to conclude that a gentle balance, not unlike the way the seasons change to relieve and succeed each other on a regular 3-month interval, was the only solution. The collective result of their determination is our Constitution, Magna Carta for the masses.

If, then, the Constitution of the United States is the keystone in our buttress against ungodly hordes who would pillage our shores like the Visigoths over the Janiculum, we must guard it with an even greater vigilance than we do all the gold in Fort Knox. At one of the most trying and pivotal turning points in America's young history, President Abraham Lincoln gave us the vital reminder of why 13 distant colonies declared their independence from a kingdom an ocean away: government of the people, by the people, for the people.

The 20th century in the United States began with consistent citizen involvement in government. Seldom was there a time when blocks of voters failed to make their leaders in the nation's capital aware of their displeasure. Sadly, that span of 100 years ended with the crushing force of a tidal wave of disinterest. Consumed with paying the bills and keeping a roof over their head, more and more voters turned away from direct involvement in the way their towns, states, regions and nation are run and focused instead on our capitalist economic system. That disengagement from politics is the greatest threat which our republic faces and the primary reason why I declare my candidacy through this document.

> *"All great civilizations which have fallen had a common characteristic: people at all levels allowed problems to pile up faster than they were solving them until the situation became overwhelming.*

> *"If we improve the general ability of people to solve problems (including their own problems and whatever problems they may encounter), we improve the prognosis of our own civilization.*

"Today's greatest problem is apathy. People no longer have an interest in larger issues which affect them. They are less informed than some distant decision maker about what would be in their best interest.

"The effect of apathy is that the people whose job it is to make the decisions which affect us all never receive your input and wouldn't care even if they did because they think they can base their decisions entirely on polls which are extremely sensitive to how they are worded. In other words, our political system does not equal our economic system. Corporations which sell to us must adhere to our wishes because we vote them out with our dollars if they fail us. The same does not hold true of our current political system.

"Democracy means that people have a meaningful say in decisions which affect them. The result is the greatest good for the greatest number. If we all have input on the decisions which affect everyone, it is more likely that the decisions will be good."

—Dr. Win Wenger

SECTION 1
Dejection And The Dejected

In Knots At 37,000 Feet

Imagine your life's greatest disappointment. Relive the jitters of awaking from a powerful nightmare. Picture yourself overwhelmed by grief. Now you know how I felt in December of 2004.

For months, I could taste the eggnog which I would share with my father, mother and sister over a roaring fire when I visited them in Maine over Christmas but when John Kerry conceded to George W. Bush, every one of my senses dimmed. Food and drink became bland and colors, opaque shades of their once vibrant hues.

There I sat, dismayed, cruising exactly 7 miles above America's eastern seaboard, unable to enjoy the plush leather seats of the Song Airlines 757 bound for Boston because nearly all of the seat-back televisions in the rows forward of mine were tuned to Fox News which ran interminable clips of George W. Bush, the antagonist of my misery.

I was in knots, overwrought with fear for the future of my country and my world. Instead, when I booked my flight early in 2004, I grinned like a Cheshire cat because with every fiber of my being I knew that I would see my sister and parents over Christmas, less than a month from John Kerry's inauguration and oh, how sweet the homemade pumpkin pie a la mode would taste!

Instead, my worst nightmare happened. 4 more years of this buffoon! How could it be? Of more immediate concern, how could I survive another 90 minutes of Bush's swaggering bravado aboard this packed aircraft? I could not impose upon 20 or so of my fellow passengers to tune their live satellite receivers to another channel. So, I set my television to the satellite tracking of our flight path, noting that the temperature outside the plane was 70° below zero Fahrenheit but then the screen flashed our current position, just inland from the mouth of the Potomac. Coming to grips with the fact that technology would not help me pass the time until we landed in Boston, I closed my eyes and pictured the chilly, snow-covered surroundings which awaited me in Bangor, along the banks of the brackish Penobscot.

Melancholy Rambling

Glimpsing the warm glow of the approaching smiles of my parents was a welcome change from the rest of my trip as I twiddled my thumbs awaiting my luggage on the carousel in icy Bangor. I inquired as to the weather forecast for my first night in Maine in many a year and shivered when I heard "Sub-zero wind chill factors." Nevertheless, I was very glad to see the familiar gaze of the people who had raised me. Somehow, I knew, though, that my brooding was caused by much more than the simple fact that the temperature in Orlando hovered around 70° F when I left.

Lest I allow my time in Maine be spoiled by events beyond my control, I engaged in a vain attempt to fill my head with childish visions of sugar plums. When that failed to bleach the stain on my soul, I did what so many had done long ago, I decided to get in touch with my roots.

Orlando had been my base since 1991 and I now consider myself a Floridian but I was just a short drive or walk from many of the places I had visited as an adolescent. So, I strolled along the snow-covered sidewalks of my teenage neighborhood and pondered. One of the first places I visited was the high school from which I graduated, on Parkway South in Brewer, which shared its parking lot with the town's largest strip mall. I thought about the field trips I had taken my junior and senior years. I remembered my teenage optimism about the future and the rosy predictions I had made about how my life would be after graduation.

Having landed in the field of computing late in my 20's, I had no reason to complain about my comfortable middle class existence as the country prepared to ring in 2005, except for the apprehension for the future, that same future which seemed so bright in 1987 as I donned my cap and gown to graduate from high school. The inner turmoil which flew with me from Florida to Maine now gnawed at me like beavers at a poplar, redirecting one of the Pine Tree State's pristine rivers.

Having suffered a youth fraught with emotional toil (much of it at my own hand), I knew that time still heals wounds. However, I knew that the scars now forming inside me were national. Literally millions of my fellow Americans felt as I did. How did I know? Many sought to relieve their pain by writing about it in cyberspace and I had read many of their postings. So, as I walked the streets of my home town which I had frequented 20 years earlier, I trembled. I shook from the cold of another harsh winter and fear for my country.

It was in that moment, on a typically sunny and frigid yuletide afternoon in Maine, Thursday, December 23, 2004, that the truth struck me like falling into a freshly plowed snow bank along State Street: if I wanted to fix my country, I had to do more.

Pain-Free Dentistry

Even though I felt pangs of Ebenezer Scrooge's bitterness that Christmas, I was very happy to be spending the most wonderful time of the year with my parents and sister. My long stroll in the snow two days earlier had left me almost speechless with laryngitis on Christmas night as the Kroehlers sat down to a home cooked turkey dinner. Nevertheless, the fresh apple cider, my favorite winter beverage, put a long-absent smile on my face and I let out a breathless "Ahhh!" and downed the last drop. I hadn't felt that good in a long time, not since I had knocked on a few hundred doors and written several dozen letters to voters around the country, encouraging all of them to vote for John Kerry.

Marc, my sister Jessica's significant other, had joined us for turkey and stuffing. He is a New York dentist who had been transplanted to Maine where he met Jessica, on the job. Even though I had almost no voice, we chatted about professional football, focusing on the Patriots, naturally, and their chances of reaching the Superbowl. After debating the team's potential in the upcoming playoffs, the topic turned to the 2004 presidential election, just as the delectable main course of our turkey dinner arrived from the kitchen. I hoarsely expressed my disbelief at John Kerry's concession and how I dreaded the thought of tolerating that idiot from Texas until January of 2009.

While I came to understand that his statement was intended to console me, Marc's words of "We only have him for 3 more years, give or take" energized me. As I digested the delicious repast which my parents had prepared lovingly, a notion began pecking its way into my consciousness, not unlike the way a hatchling frees itself from the egg which its mother keeps warm with her feathers. I was and remain powerless to do anything to rid America of George W. Bush but if I wanted to prevent another of his stripe from ascending, I needed to start immediately. In short, the key to relieving my bitter sorrow lay not in the present but a strategic focus on the future.

Even before I departed for Orlando, though, I had a bracing reminder of just how great the damage is in the present. As I awaited the announcement of my flight departure at Bangor International Airport, I shared the spacious staging area with hundreds of uniformed Army reservists on their way to the Middle East. Seeing so many squared jaws, laced combat boots and misty relatives who prized every second of togetherness before their loved ones descended to the tarmac, I was almost paralyzed with grief. If their mission was accomplished 18 months earlier, why were we sending hundreds of fresh soldiers to fight? Never in my life had I seen such a stark reminder of the human cost of war and the very real consequences of bad policy.

The Happiest Place On Earth

My wife, Catrin, knew, of course, about the countless hours I had volunteered to help send John Kerry to the White House. She also keenly recalled the look of stunned dismay which filled my face when Senator Kerry conceded the election that previous November. She did not, however, know the degree to which a week-long visit to Maine had filled me with a fresh sense of optimism. My 3,000-mile journey was more than a chance to hug the folks. It was more than a chance to sing in a family quartet on Christmas Eve (albeit without my usual baritone mellifluence). No, my tormented yet delightful holiday trip was the cipher code to the first third of my life and the prologue to the next. I was a (still largely voiceless) new man and had the toothy smile to prove it.

I love fireworks. Some might say that I am a connoisseur of nighttime chemical illumination. Living just 30 minutes from Walt Disney World is wonderful but when I returned to Orlando to celebrate the new year with Catrin, the happiest place on earth was our home, just the two of us and our four cats.

The Deluge

Catrin and I tried to remain awake to watch the ball drop in Times Square on ABC's Rockin' New Year's Eve but a heavy downpour made it impossible for us early birds. However, the torrential rain was a portent of the wave after wave of thoughts and feelings which would occupy my mind for the first half of 2005. What began as an unusual weather event became the beginning of insight and the successor milestone to Marc's words of consolation expressed on Christmas night. Each raindrop represented a glimpse of the future. Together, ripples of insight became waves which swept me away, although I did not know my destination when I brought in the first edition of 2005 of the local newspaper.

All too soon, my winter vacation ended and it was time to return to my job as a web developer. I managed to focus on my work better than I had in the days between the election and my trip to Maine. The fresh optimism coursing through me provided newfound energy and I focused on each task which crossed my desk safe in the knowledge that I would reach a crossroads very soon. Sure enough, I reached that intersection sooner than expected.

Because of my work on the Executive Committee of the Central Florida group of the Sierra Club, the nation's oldest and largest grassroots environmental advocacy organization, I was invited to the ground breaking of Florida's Hydrogen Highway, just a few short miles from home near Orlando International Airport. When I read the invitation, my eyes paused understandably on the line which explained that Jeb Bush, governor of Florida at the time and brother to George W., was to be the host and keynote speaker. I summoned my courage and submitted my confirmation to the appropriate office.

As time passed and the day of the ground breaking drew nigh, I was pleasantly surprised that the butterflies in my stomach did not turn to wasps as I had feared. Indeed, I felt a sense of guarded hope that Governor Bush intended to model aspects of Florida's Hydrogen Highway after what Governor Schwarzenegger had done in California. When I arrived at the unimproved, mulched lot along Boggy Creek Road, I was pleased to see several prototype hydrogen vehicles already on display. As I collected my event program at the registration table and pivoted 360° on the aromatic pine mulch, I half expected to see some indication that the ceremony would consist of an excess of pomp and little substance. I was in for a pleasant surprise.

Lighter Than Air

Imagine the perfect day for making history: breezy, sunny, with temperatures around 70° F. That was the weather on February 18, 2005: perfect. Just a few high, wispy clouds caused the most random of breaks in the almost blinding sunshine of south Orlando. I knew that mother nature was smiling on us for participating in the auspicious ceremony to herald Florida's first institutional foray into alternative energy.

January and February offer arguably the best weather for snow birds in Central Florida. Even though I am very proud of the fact that I live in the Sunshine State year round and have done so since 1991, I couldn't help but be glad for Bill Ford, Jr., one of the day's speakers, because the conditions at the outdoor venue had to feel welcoming to a Detroit dignitary and captain of industry. Oh, what a Friday was in store!

Despite the irrefragable malaise which gripped my soul's perennial optimism, a sense of heady glee fell upon me as the Governor's motorcade arrived with Mr. Ford seated alongside Mr. Bush. Orlando Mayor Buddy Dyer led the front row assembly of other dignitaries in greeting the honored speakers and the event began. Within a few short minutes, I knew that I was present at an historic event. The bluster which one expects when bureaucrats, executives and politicians gather on the same stage to praise each other was replaced by exaltation for the future of alternative energy, with hydrogen and solar power at the forefront. The only hot air present occurred naturally.

Each of the event's speakers told us how important it was for Florida and the nation to embrace renewable sources of energy in order to strengthen our economy and protect our natural environment. From Chevron Research to Ford Motor Company to Progress Energy, the audience heard from distinguished gentlemen who very easily could have occupied their time elsewhere. They chose, instead, to lead by example and everyone gathered on that glorious morning benefited from their collective zeal for hydrogen.

I felt like a kid again, inspired by the possibilities.

Absence Of Swagger

As each speaker concluded his remarks and received a gracious round of applause from the enthusiastic audience, my cynicism compounded itself until I half suspected that what I was hearing was all too easy—rhetoric the color of chlorophyll but policies black as coal. When Mr. Bush and Mr. Ford took time to address a flurry of questions from the press, however, my exuberance returned.

In addition to providing an atmosphere of old Florida in its natural state, the mulch lot where the event was held allowed for very informal gatherings after the staged speeches had ended. In short, I was able to listen to the conversations which Mr. Bush and Mr. Ford had with the reporters and draw my own conclusions.

Because of my work on energy issues with the Central Florida Sierra Club, I already knew much of what Ford Motor Company and Governor Bush had planned with regard to hydrogen and renewable energy so I took the rare opportunity of examining the characteristics of each man's answers in a way which ordinarily is unavailable to me, the body language and the unspoken answers. I was surprised by my findings.

I had been very disappointed with Bill Ford, Jr. because he had broken many environmental commitments which he made to the public before ascending to CEO. His company had the chance to put the country's auto industry on the right track yet decided instead to fight the environmental community, including the Sierra Club, at almost every turn. This profound sense of disappointment which I felt led me to watch his television appearances with keen awareness. Yet, when the same man stood not 10 feet from me, fielding questions from the local and regional press, I questioned my recollections.

Bill Ford, Jr. was not only congenial but quite accessible, to the point of being humble. He did not strut about like the king of one of the nation's largest empires. He did not lift his nose above his eye line. He did not dismiss any questions posed to him as beneath his response. Simply put, he did not swagger. I had met no billionaires or chiefs of billion-dollar behemoths until that moment and consequently was struck by the humility I perceived.

I found Governor Bush's reserve and lack of pretense equally refreshing. Here was a man whose brother occupies the White House and whose father was a war hero, respected former member of the United States Congress, past Director of Central Intelligence and, of course, the 41st President of the United States yet he conducted himself like any other Floridian. Since I had disagreed with many of his decisions and policies, I couldn't bring myself to say that I was impressed but his demeanor was memorable for its polish and sincerity.

I expected to find traces of brother George but instead felt like I was watching his father during the time of the 1980 Republican national convention—advertent.

Only Wonks Need Apply

My daily tear-off calendar changed all too slowly from Monday to Tuesday to Wednesday as I pondered the larger meaning of what I had seen at the ground breaking of Florida's Hydrogen Highway. Was there no way for me bring my passion and zeal to bear in order that America might end its addiction to petroleum and its servitude to OPEC? Surely, if my colleagues in the Sierra Club were satisfied with my knowledge and abilities to serve as an advocate for Mother Nature, I was up to the task of doing more. But what?

I enjoyed following national and world politics and the event with Governor Bush had given me first-hand knowledge of how happenings hundreds or thousands of miles from my home can have lasting, local impacts. How could I inject myself appropriately? How could a novice jump from private sector work as a web developer to affecting public policy?

"Run for office!" Excuse me? *"Run for office!"* Dr. Howard Dean, former Governor of Vermont and new Chairman of the Democratic National Committee asked all the people who had supported him and volunteered for his presidential campaign in 2003 and 2004 to consider running for office if they were unhappy with the results of the 2004 election. When I had received a similar exhortation from the coordinator of the local office of MoveOn.org where I had volunteered in September and October of that year, I dismissed the calling as desperate. I am nobody. My special abilities lie in writing computer code, proofreading the writings of others and critiquing contemporary motion pictures. Run for office? Oh, please!

However, I had followed Governor Dean's campaign for the DNC chairmanship with keen interest because his presidential campaign had constituted a genuine insurgency in its early months. What's more, because I work in the web development field, I had remembered how historic his candidacy was during the 2003 MoveOn.org virtual caucus—the response from Democrats all over the country was so strong that the website became overloaded. Governor Dean had his head on straight and knew a few things about being a political outsider. If he wanted me (patriots who otherwise just knocked on doors, signed petitions and gave financial donations, that is) to run for office, it must be a good idea and something which big oil's special interest groups would like to avoid at all costs.

Which office, though? For which office should I run? Well, what are my skills and what are my qualifications? I enjoy the theater of national and world events. I know that most of our nation's landmark environmental gains have occurred or at least started at the federal level. I am very patriotic. I believe that greater Orlando is America's new melting pot for the 21st century. I think globally and act locally. So, what's the answer?

It did not come to me right away. Indeed, the wheels of my imagination turned rather slowly on the matter of running for elected office, until late summer. I was pleased and much honored that the Central Florida Sierra Club saw fit to subsidize my attendance at the Solar World Congress, which took place in August just a few short miles from my home. At the landmark global event at the Rosen Centre on International Drive, I attended session after session headlined by many of the greatest minds in the field of alternative and renewable energy the world over. I was awed by their knowledge. I was inspired by how revolutionary their technologies were. I knew that I had found a home even if chemistry and physics weren't my strongest suits. If that weren't enough, each successive presentation which I saw at the multi-day event ran round and round inside my head as I connected it to legislation which had passed or never even received a vote in Washington. If I wanted to help the people who had taken the time to come to Orlando in order to share their knowledge about the means to break the cycle of petroleum addiction, I need to work in the nation's capital.

Then, the light bulb which had snapped on in my head was replaced with a carbon arc spotlight from the moderator's table at the final session of the gathering. Mrs. Sue Rolf of the historic city of Oxford in the United Kingdom spoke about the upcoming International Solar Cities Congress to be held in six months' time. Mrs. Rolf's English was beckoning and the urgency of the need to keep the fire of optimism at the Rosen Centre burning bright quite compelling but travel to the United Kingdom? I've never even been there! By the time she had concluded her presentation, though, I was ready to book my flight.

Although most of the presenters whom I heard during the event on International Drive were not from the United States, I had, either deliberately or by natural inclination, drawn conclusions between most challenges explained and their connection to American policy. The invitation to travel to Oxford for the International Solar Cities Congress was different, however. It was the culmination of a dream which had begun because of the influence of my countrymen. Mayors from several American cities had committed to adopting the climate change restrictions of the Kyoto Protocol. The dedication of those courageous public servants to effecting change in spite of the well-funded opposition and global warming critics in Washington led Mrs. Rolf to seek to do more was inspiring. It was good old Yankee ingenuity which inspired an Englishwoman to extend a transatlantic olive branch during a time of extreme pessimism as to the viability of any meaningful accord to reverse the effects of global warming. If big city mayors, some of them Republicans, could commit the equivalent of political heresy and affirm to the whole world that humanity is the chief cause of climate change and that reforms must begin immediately, then a (sometimes) humble, (mostly) unknown environmentalist from Orlando has a chance to do his part.

"So, where do we begin?"

My inner voice was not swayed by all of the inspiring flashes of the future which I had fed it. I had to connect the dots from a to b to c in order have a real plan of attack, even if the only doubter I faced at that time was my own subconscious mind. How about, instead of a political campaign, a more formal version of the volunteer work I currently did? I already was familiar with the inner workings of the Sierra Club, especially its successes in grassroots lobbying of Congress and the White House on a range of environmental issues. However, because I had donated time and money to other green organizations over the years, including the Environmental Defense Fund and the Natural Resources Defense Council, among others, I knew that structured advocacy could be quite rewarding. The prospect of working as a Washington lobbyist didn't sit well with me, though. Catrin and I would have to move to the Potomac basin or I would have to commute, an arrangement which would mean a great deal of time away from home.

As I pondered every option which came to mind of just how I could affect change on Capitol Hill yet still keep my roots in Central Florida, the real answer became increasingly clear. Governor Dean was right. Elected office was the only choice.

"You, a politician? Don't be absurd!" I tried again in vain to devise a viable alternative. I attempted to talk myself out of the whole matter. The harder I tried, though, the more firm my conviction became. I simply cared too much about the future of my country and my planet to be an idle bystander. The time to involve myself is now and the place is Capitol Hill.

"OK, great, Mr. Politician! If you want to stride the hallowed halls of Congress and help save the planet from fossil fuels, you have to convince a great many Floridians to take a chance on you and send you to Washington as their proxy. How many Floridians do you know well enough to ask for their vote?"

My inner voice had a point. If I meant to do this, I needed a plan and an internal assurance that I had my head on straight. Running for Congress would mean an enormous sacrifice and vast dedication of time and resources, not to mention the need to raise a great deal of money. How would I go about this and what were my chances? I thought a great deal about Governor Dean's insurgent campaign and Senator Kerry's victory in Iowa and New Hampshire during the 2004 primaries. In the end, though, my first moment of clarity as I discussed my future campaign with myself was a very important statement by Senator Hillary Rodham Clinton. When asked whether she would run for the White House in 2004, she stated that she intended to serve her entire 6-year term as a Senator. She went on, of course, to win reelection by a very comfortable margin. However, the relevance of her 2004 decision on my decision to run for Congress in 2008 was her candor: she said that the voters of New York took a chance on her in 2000 and she owed them. It was her duty to serve 6 years as a Senator to the best of her ability.

During my one-third century on earth, I learned what it meant to be a polarizing figure. In school, my classmates either loved me or hated me. Few were indifferent. The same was true of colleagues during my professional life. Rarely did I evoke a lukewarm reaction. Hence, if a figure as well-known as former First Lady Hillary Clinton felt honor-bound to eschew political expediency and live up to not just the letter but the spirit of her obligation, then I had a responsibility to apply the knowledge I had obtained as a Sierra Club volunteer to the larger battle against climate change. Besides, if I planned to fly all the way to the United Kingdom to hear what Oxford, a city roughly the same size as Orlando, though much older, was doing to decrease its production of greenhouse gases, I needed to have a plan in place so that I didn't leave there powerless to respond to a dire situation as it would be laid out for me. Even if I could do nothing right away, I knew that I would feel foolish to travel such a great distance, spend more of the Sierra Club's limited resources and leave without a plan of action.

So, I set a task for myself. There are two houses in the United States Congress. I needed to figure out which was the right place for me. I needed to figure why it was the right place and I needed to devise a strategy of being elected.

The first step was quite simple. I have only a small group of friends who could assist with my campaign and most of them live in Central Florida. What's more, I have no campaign war chest which means that I would have to continue working as a web developer in order to make my monthly mortgage payment. If I intended to do this, it would have to be an operation based in Orlando and targeted at Orlando. So, if I couldn't operate statewide, I couldn't run for the Senate. If I had two possible offices to pursue and one of them was impractical, two minus one equaled one, the United States House of Representatives.

Florida's eighth congressional district it was!

"Uh, what do you know about the eighth congressional district?" My inner voice sure could be a wet blanket at times! I still tingled from the conclusion that I was running for the House of Representatives and it wanted to shift back into strategy mode. I sighed, "No rest for the weary, I suppose, even if my debate opponent is myself!"

Florida's eighth congressional district had been represented by Republicans for as long as I could remember. What's more, it was my Congressman who had led the effort to impeach President Clinton so I knew that my wishes had not been heeded for quite some time. The key point, though, was why my vote for the Democratic challenger in 2000, 2002 and 2004 had not resulted in victory for the challenger. That crucial detail would have to be included in my strategy for winning the seat.

I now knew how to begin to create the first outline of my campaign gambits and tactics but I was a long way from a viable plan. My mental process became chronically bogged down on the point of electability. I knew that I had the environmental credentials and more than enough patriotism but how could I persuade potential constituents to gamble on a novice? The answer was my wonkish way.

Because the 2004 nomination process for President was so important to me, I used the Internet for the first time to research the Democratic candidates. I watched their speeches. I read their interviews. I perused discussion boards and blog postings. I wanted to learn all I could in order to make the right decision. Without realizing it at the time, the process of forming opinions about each challenger yielded the ability to express an opinion about most any political issue based on my values as a true blue Democrat. Moreover, since I had developed persuasive abilities to convince people to sign petitions and pay membership dues as a Sierra Club volunteer, what I needed to do was combine the two.

The solution came in realizing that Florida's closed primary system means that I only have to communicate with Democrats for the first part of my campaign. I only have to worry about convincing fellow Democrats that I am the right choice to face the Republican nominee in the general election. Environmental credentials plus strong Democratic beliefs multiplied by opinions regarding current events equal a genuine candidate!

Start With The Base And Work Outward

Liberty! Sweet relief! I felt like I had just passed the road test and obtained my driver license! The first draft of the first section of my campaign was written and I liked it! I had used my directional indicator and trailed my brakes every time I approached a mental obstacle and I still was smiling. It must be all downhill from here!

In theory, yes, it was all downhill. In the comfortable sterility of my mind, a plus b must lead to c, victory in this case. In the real world of smile-and-shake campaigning, I still had a lot to learn. The biggest question was, where to begin?

"Well, how do other first-time candidates begin their campaign?" They announce themselves through broadcast media and direct mail. I am starting from scratch with only a set of ideas so that won't work. What else?

"They ask their friends and family to knock on doors for them." Now, there's a notion. The Sierra Club! Legally, I cannot ask for sanctioned support without an endorsement from the organization but its members are not just potential voters, they are my friends! Sure enough, as I mentioned my campaign to just a few of them, their eyes lit up with excitement. In one case, I saw goose bumps rise on the person's forearm. I was on to something!

Within the span of just one evening, the first stage of my retail efforts was born; I'll start with what I know, the environment, and rally those who know me and for whom protecting our natural environment is priority 1. I will harness their zeal and convert it from a ripple into a growing wave of support.

Face Time

What is the best way of channeling grassroots energy and focusing it on people who otherwise would not be involved? The Internet! Governor Dean proved that. The answer must lie there!

One of the reasons I have been very happy working as a web developer for many years is that it is a highly democratic medium. It was invented with American tax dollars and it was released to humanity as the information superhighway. Just as a President had grasped the importance of connectivity between citizens when Dwight Eisenhower fomented the interstate highway system, President Bill Clinton knew that he and Vice President Al Gore were on to something when they saw to it that all public schools should have Internet access, even wiring some of them personally.

The year 2006 saw the rise of what is known within the Internet community as Web 2.0, the true emergence of interactivity between individuals in such a way that their conversations may be shared for all the world to see. What better way to integrate this new form of one-to-one-to-many contact than through the boiler room of retail politics! In short, it's only a matter of phrasing the message properly, in a disciplined manner and then launching it into cyberspace.

2006 also taught us that the Internet is a vital tool in organizing and persuading but not necessarily in assuring that supporters take the time to cast their votes on a given Tuesday. This process must be achieved through America's great and lasting tradition of retail politics, of knocking on doors and leading house parties. Put another way, exposure is key, especially for an unknown candidate.

How, then, does one convince strangers to welcome him into their homes along with several other strangers? Issues! There is more than enough selection of what is wrong with America today from which to choose and motivate people. Besides, starting with the base means that the first groups will be comprised of only Democrats, making us all fast friends. Issues are the key and rallying Democrats to decry the reality of those issues can start the ball rolling, rolling like water, which becomes a ripple, which can merge with other ripples to form a wave.

Blue Dog In A Red State

The statistics are clear. Florida is a blue state, when voter registration rolls are measured in aggregate. The eighth congressional district is no different. Sadly, though, many times blue precincts, counties and districts vote Republican, sometimes because they don't see a familiar Democrat on the ticket and other times because change feels dangerous.

Our party, though, now is winning on the basis of change because the status quo represents disaster. As important as such a concept is, we must do more. My candidacy stands for just that. Fiscal responsibility is the best place to begin and there is a group of Democrats in Congress who champion the need to spend the people's money efficiently and without reliance on long-term deficits: the Blue Dog Coalition. Their combined voice can and will be a clarion call in 2007 and on into 2008 and if the voters of my congressional district do me the honor of sending me to Washington as their representative in the House, I will add my resonant voice to the Blue Dogs. Anything less is fiscally irresponsible.

Allow me to proceed, then, dear reader, to outline the crucial challenges and themes which face all Americans, especially my fellow Floridians. Indulge me by freeing your mind to follow my conclusions and remedies for what ails our political system. Trust me to lay out a new progressive pathway, brick by brick, a pathway which leads to an all-inclusive America, a nation which is genuinely truthful and just.

Together, let's glimpse the future. Join me as citizens from coast to coast who feel left behind by a political system which only recognizes candidates with deep marketing budgets take a collective peek at 2008. It is the beginning of the end (of big money in politics).

SECTION 1
Toxins And Transplants

Surrender In Paradise

Only twice in my life have I felt lighter than air. The first time was body surfing on the Jersey shore during summer break from elementary school. I attempted to catch the perfect wave many times when 12 and 14 tender years of age but it was the very first time the tide lifted me toward the sky which lingers. The sensation was positively unique.

The second time was during my tour of the University of Miami in February of the year 1986, during my junior year at Brewer High School. When my father and I left Maine, daytime temperatures had been below freezing for weeks. Touching down at Miami International was a profound shock both atmospherically and culturally. My first, brief stay in the Sunshine State left a lasting impression, so much so that I knew that I wanted to be a Hurricane after graduating high school. Miami was paradise and I had surrendered to it like a wild horse which finds comfort in a busy pasture.

After the obvious difference in temperature, the intensity of the sun's bright, burning rays was the characteristic I noticed most. Before visiting Miami, never in my life had an opportunity presented itself to test what I had learned in earth science class. Being almost 2,000 miles closer to the equator really did make a difference in the weather!

The most lasting memory and the event which convinced every fiber of my being that Florida is the home of my destiny was driving to the beach along Interstate 195, the Tuttle Causeway, in our rental car. Back home in Maine, one simply does not open the window in February. Frostbite is a fleet-footed enemy. However, there we were, at dusk, traveling east with all of the windows open, fixated on the breathtaking pastel hues of the darkening sky and the caress of the tropical breeze on our face and through our billowing clothes.

In 1988, when I decided to leave Miami to attend Full Sail Center For The Recording Arts in Orlando, I found that I liked Central Florida even better. I knew that I would miss the Atlantic Ocean and South Beach but the wide-eyed child still alive and well inside me could not help but love being less than an hour's drive from Walt Disney World. The cloud formations which filled the sky weren't quite so lovely as in South Florida but, ah, the sun! Vermont may be known as the Green Mountain state but Florida should be known as the Green state because it seems to be in full bloom every day of the year!

After my first decade of life in Florida, the term global warming became somewhat common parlance on broadcast news and I started to take closer notice of things. What once was a sultry paradise seemed to be losing some of its subtropical moisture. Street corners which had green grass no matter when I passed them had become brittle like the crust of day-old toast. Golf courses, not surprisingly, were lush no matter when I glanced at them but I began to fret about our overreliance on the aquifer for irrigation and the effects which climate change would have on our ability to provide potable water for the state's bulging population. Not only were we drawing ever larger quantities from subterraneous flows but Mother Nature seemed to have reduced our daily allowance.

When I moved to Orlando from Miami, I was amazed at the daily downpours, so regular during the warmer half of the year that I didn't need to wear my wristwatch during the afternoon. If it had started to rain, the time clearly was 3:30. I knew not to make outdoor plans before 5:30. That's all there was to it. Yet, as we approached the new millennium, I didn't find that planning around tropical weather patterns was quite so necessary anymore. Global warming had my home in its grip. I worried but did not really think that there was much I could do about it besides driving less, replacing my home's light bulbs with compact fluorescents and running the climate control system less.

Suddenly, George W. Bush happened. He told us that adopting the Kyoto Protocol was impossible without damaging the American economy. My growing interest in the science of climate change and the results of burning fossil fuels informed me otherwise. Since Mr. Bush's father was the man who committed the United States to Kyoto, I couldn't bring myself to believe that George, Jr. would break his campaign promise to regulate carbon monoxide and not offer an alternative measure. Sadly, as expert prediction after prediction about global warming came true, George's response essentially was, "Everybody should drive an SUV."

The reaction by my countrymen varied from condemnation to indifference, akin to saying, "If George W. Bush says that nobody knows that global warming is real, well, I guess I don't have to worry about it. Besides, sport utility vehicles are fun to drive." I knew the facts and remembered that our broadcast media had done a passable job of explaining the causes and effects of climate change when Bill Clinton was in the White House so what was the problem? Why weren't there protestors marching in the streets of every major city? After all, the scientific community was unanimous on the matter. No answer presented itself so I went on fretting.

By the time Hurricane Katrina struck south Florida and then smashed into the Gulf Coast a few days later bringing mortal terror to an entire region, the debate had ended. We no longer had people on the fence about climate change. The cacophony of voices clamoring that more study was needed or that dissent existed between botanists, climatologists and geologists had settled into just a pair of choirs singing in unison. Choir one chanted, "Global warming merely is the result of a natural cycle of temperature rise which occurs every 10,000 years." while choir two was led by President Clinton and backed by esteemed scientists such as Sir David King in chanting that global warming is a greater threat to the United States than terrorism.

It is one thing to hear the bad news on television or in my local Cineplex in the case of "An Inconvenient Truth." It is quite another to attend an international symposium on global warming, as I did one month before Katrina in Orlando at the Solar World Congress and just a few months later at the International Solar Cities Congress, and have experts tell it to my face. As I sat in the austere yet ornate testing school of Oxford University, the irony of seeing just how much of Florida would be lost to permanent sea rise over the next decade was not lost on me. Moreover, despite being on another continent, 4,000 miles from Florida, when I heard the devastating predictions and saw how maps of my home state would have to be redrawn, a burning urgency to call upon every American to help protect Florida from sea water welled up in me like a geyser.

I sat quietly, admiring the portraits of headmasters who had administered exacting tests to thousands of British pupils in centuries past, not knowing how or where to direct my angst. Florida was facing its greatest offshore menace since the Cuban missile crisis and there was little I could do about it.

The next week, on the flight home to America, I had a window seat. I watched the Welsh coast disappear in the distance as we climbed to our cruising altitude. Even though the closest I ever came to meeting a person from Wales was watching Catherine Zeta-Jones in "Chicago," I felt a tingle of kinship to the people living below. The existence of their home was in danger every bit as much as mine was. Worse still, perhaps, they had little or no voice in the American political system, the system which encouraged its people to consume fossil fuels recklessly, the system which blocked efforts by its allies to deal with climate change while we still have a chance to avoid the most damaging consequences. I had to do something!

By this point, early April of 2006, my decision to run for Congress was final. Now, my thought process had to turn to issues and strategy. I knew that the incumbent, Ric Keller, was a rubber-stamping Bush team player so he clearly supported our continued reliance on fossil fuels. No problem there. However, I also knew that a winning strategy cannot be based on simply not being one's opponent. Indeed, in the 2004 election, Senator Kerry won Orange County, of which Orlando is the seat, by a comfortable margin yet the Republican incumbent Congressman whose district covers much of Orange was reelected. Why?

I know my community. I know its mores and I know its political persuasions. Orange County voters are predominantly Democrats. However, if that is so, why has the eighth congressional district been represented by a Republican for as long as I can remember? The answer is apathy. Inadequate numbers of Democrats take the time to vote by absentee ballot, to vote early in person at the Supervisor of Elections office or to vote at their local precinct on election day. In other words, Republicans are winning by default.

Clearly, my conclusion failed to included independent voters, whom Florida classifies as NPA (no party affiliation), but polls indicate that they vote for Democrats more often than Republicans given a clear choice. When assessing the situation strictly between the two main parties, though, apathy is the explanation.

Apathy, Apathy, Apathy

The word kept repeating in my head. I was disappointed as a Democrat and as a patriot that so few American voters and particularly so few of my fellow Central Floridians participate in the voting process. I pondered the matter for days and days before tiring of my quest for a solution to the problem. Then, as so often occurs in my life, the solution did not present itself but a catalyst to the needed thought process did: the country learned from the world of television that more citizens vote for their favorite contender on "American Idol" than in political elections. I chuckled at the irony but at the same time wept inside for the future of the republic which I hold so dear.

How could it be that people pay more attention to the results of a television show than how their government is elected? I knew that public school budget cuts had slashed the time which high school students spend learning civics and how their government works but surely they still understood that elections are about choosing how a person whose salary comes from their taxes will vote in Washington! Yes? Maybe not!

I brooded for days after hearing of the alarming statistic about the franchisement of voters to "American Idol." I genuinely was at a loss as to how to process my emotions. I knew that I was embarrassed and discouraged by the news but could see no further. The more I dwelled on it, though, the more my thoughts drifted to the statement it made about my culture and society. If America is the economic leader of the world, then surely other nations will copy our flaws and virtues with equal vigor. When I visited England for the International Solar Cities Congress, I saw how Americanized many things were. Indeed, Starbucks coffee shops were more common in London than Orlando! The more I brooded and the more I pondered, the faster my mind circled the same point. I felt like I was ten years old again, playing a solo game of tether ball. I could smack the orb but doing so yielded no entertaining results.

Fixation and single-mindedness ultimately yielded fruit when I learned the statistic that 2005 was yet another record year for global high temperatures. We are in trouble! As tragic and infuriating as the announcement was, it had a benefit: my mind stopped dwelling on voter apathy for a moment. I now could worry about emaciated polar bears whose homes on large ice flows had melted and crumbled into the Arctic Ocean and the island nation of Tuvalu, on track to be the first sovereign country to disappear because of rising seas.

Apathy, Global Warming, Apathy, Global Warming, Apathy, Global Warming

My mind had become a roulette wheel, alternating between bad and worse. I generally am an optimistic person who doesn't allow a rough bout to lead to dismay. The prospects for America's immediate future had soured to such an extent, however, that I couldn't help but feel blue. As I wallowed, I wondered how Florida would weather the impending storms of a detached electorate and an unrelenting sea poised to bring the Atlantic Ocean several feel closer to home than it was the last time I checked. Then, suddenly, without warning or fanfare, my mental light bulb switched on. The first step to solving the challenges of global warming and voter apathy is the same, political innovation, innovation which is radical yet seems incremental. My candidacy had to be part of the solution and addressing my two top issues had to be at the heart of everything I do.

Americans have disengaged from the political process because it loses its relevance to them. Why? Well, there may be a hot button issue or two which can energize small blocks of voters but when people see rampant systemic failure, they lose hope. They see huge gaps in our health care safety net, so big that an entire generation has entered the work force with inadequate preventive care. They see an economic system which places homeownership out of reach of single-income families. As a result, young parents find that both they and their spouse have to work and their children are raised by three parents, the third being daycare.

Potential voters feel detached from their government when they see homelessness on the rise again, morphed from a simple question of basic services and human dignity to a political hot potato with warnings of the dangers of a welfare state. Perhaps worst of all, they withdraw behind the windows of their apartments and cars when the word Christian becomes a definition of political party rather than faith.

The good news, though, is that the solution is within our reach. The challenges we face and the cause of such profound discontent among our citizens merely are a new variation of the themes played out in America's fledgling years. By studying Thomas Paine and the Federalist Papers, I learned that our earliest patriots faced similar struggles. Their response to those struggles was to declare our independence from Great Britain and later to help convince a timorous new nation that the insightfully written Constitution which had been introduced for ratification was the ideal compromise, that its checks and balances were both intended to smooth rough edges and prevent despotism.

In short, in order to help rekindle America's flame, the passion for politics in the public square which produced such rhetorical standards as the Lincoln-Douglas debates and Senate filibuster must return; we must take our system back to its roots. The mainframe of our democracy which is our two-party system in all its intricacies must be rebooted with a new operating system. It is time to set higher standards. As soon as average Americans find that our political system is useful and sustainable again, they will engage it.

"When facing the future there are two choices: slow death, or deep change."

—*Robert Quinn*

The structure of our federal government, from lifetime appointments of judges to 4-year terms for presidents, to the manner in which spending bills are introduced in the House of Representatives but ambassadors are confirmed by the Senate, is by astute design. The Founding Fathers knew what had worked well for the Romans, the French, the Germans and others. They also knew what had failed those groups as well as, of course, the flaws of the British parliamentary monarchy. They took the best, discarded the worst and then debated the rest. They knew that with courage, faith and perseverance,

America could endure and flourish. For its first 200 years, except for some painful missteps, flourish it did.

As the United States neared the end of the 20th Century, however, standard procedures which had served it so well for so long failed to keep pace with societal changes, especially the manner in which Americans earn a living and learn of current events. Television, once little more than a novelty, now looked after latchkey kids and replaced conversation during dinner. Suddenly, candidates who could raise large sums of money didn't have to worry about knocking on doors so much as controlling television since many voters prefer to watch TV than answer their doorbell. Where once a person's ability to rally crowds to their feet in support was the best measure of a candidate's viability, now the degree to which he or she is photogenic or telegenic tended to be most accurate.

The checks and balances between the three branches of our federal government do more than set forth who is responsible for what and who supervises whom. They also resemble a human pair of lungs. When collective passions run high or a group of officials, either appointed or elected, are swept up in political furor, the branch which oversees the affected branch can intervene to restore dispassion, to encourage those affected to take a collective deep breath. The aforementioned disengagement of the electorate from the political process and the growing popularity of television as the news medium of choice for most people acted like carcinoma, a cancerous growth which begins as a tiny blemish on an otherwise healthy organ and grows until the disease is systemic. America's political lungs had developed cancer and the replacement of substance with style by its leaders was gnawing away at her thoracic cavity.

Oftentimes, the treatment regimens for cancer can be divided into two types: remove the affected tissue surgically or flood it with irradiated chemicals. I am no physician but I am certain that our republic needs all of its organs in order to remain healthy. The answer is a transplant. In this case, though, we should use surplus material from elsewhere in the body rather than taking it from another body. In other words, given proper guidance, America can treat herself.

SECTION 3
Nonnegotiables

Vapid And Breathless

During the period of roughly seven years between the first headlines about Monica Lewinsky and Dick "Fortunate Son" Cheney's statements that the Iraqi insurgency was in the last throes, the American political system had the life nearly beaten out of it. At times, I wondered how much Beltway bitterness the average taxpayer could or would tolerate. I knew that life overall was better under President Clinton than it had been in quite some time but still had a sense of foreboding about the future. In my interactions with colleagues, friends and families, I had a strong sense that political zeal was a very rare commodity. The republic had become vapid.

When Al Gore conceded to George W. Bush by deciding not to persevere in his legal attempts to prove that he had, in fact, won the White House, I reminded myself of the things I liked about George H.W. Bush's tenure in the executive branch. Watching his son on the national news as he prepared for his inauguration, I told myself that familial loyalty would prevent him from taking any radical steps. I tried to convince myself that the elder Bush's many foreign policy accomplishments would serve as a template for Governor Bush to build what President Clinton called "A bridge to the 21st Century." History demonstrated just how foolish my internal consolations were. George W. Bush had done such an awful job as chief executive that it would be difficult for any of his successors to be considered worse.

Not one to languish in despair for longer than necessary, I tried to accentuate the positive and focus on the future. Bush would not be in the White House forever and repairing the trail of damage left in his wake would require a great deal of time and effort by great people. Besides, George W. Bush had made some good decisions, some surprisingly good, such as creating the Northwestern Hawaiian Islands Marine National Monument. I forced myself to remember that as bad as matters were, there remained cause for hope. Such hope evolved into resolve and resolve into a mission statement: the best way to set America back on course was to pick up where the Clinton presidency ended, to resume that important work, in order to raise quality of life for everyone.

President Clinton taught us many lessons during his two terms in office but few if any of them would be considered radical by reasonable people. Moving the poor from welfare to work was an important goal which taxpayers can support. Likewise, the quest for peace between Sinn Fein and Northern Ireland or Israel and Palestine benefits not just their respective geographic regions but the whole world. Peace is good for humanity. In short, Mr. Clinton showed us that some situations called for intrepidity while others demanded cool, calm deliberation and perseverance. If I wished to rejuvenate my country's political system and help it catch its breath, I must start with an idea which is reasonable to even minded people but radical to the establishment.

Where should I start, though? Whose achievements could serve as a model? The Reverend Doctor Martin Luther King, Junior certainly accomplished more than most and with history-making ramifications. I never was oppressed as he and many of his followers were, though. Perhaps I should examine the business world, especially since George W. Bush promised during his 2000 campaign to run the national government like a business. Well, as a Democrat, I knew that I would have to tread lightly since many businesses have tried to rise to greatness on the backs of their workers.

Aha! I could look to people who succeeded in business while holding true to populist ideals. Michael Bloomberg, now mayor of New York City, was not a perfect example but his company had become a standard bearer in the world of finance. What did he have to say about innovation?

"Take an established industry with entrenched players and bring in a new kid with drive and passion and he/she will rewrite the business."

Drive and passion? Eureka! I had those in abundance. I know what we need to do in order to grapple with global warming before Central Florida becomes oceanfront property and I know what is wrong with America's political system. I should begin there!

Few moments of greatness have been achieved without a plan. I could not escape that simple but compelling truth. I thought about people who had tried to change our political system. I thought of Ross Perot, for whom I had voted in 1992. With a strong sense of reverie, I remembered the things he had done right and the conclusions I had drawn about why he failed to win the White House. Naturally, my thoughts turned from those of my first days of familiarity with President-elect Clinton to words he uttered after leaving office, as he reminded

Democratic supports of the key to victory, "Success in politics depends on long-term planning and a strong, early start."

So, I needed a plan, more specifically, a good plan and I needed a way of propagating my convictions about how we bring about true reform. How did Dr. King begin? He was a popular but relatively unknown preacher from the South. Surely, if I emulated some of his strategies, I could obtain some of his results, especially since I would not have to fight riveted prejudices because of the color of my skin.

With the decision made to draw inspiration from Martin Luther King, I had to choose from the vast archive of wisdom which he had left us in the form of speeches and writings. The battle I was about to begin was essential for the health of my homeland's democracy but it wasn't about race per se. Ultimately, I landed on one of Dr. King's sermons. He gave it just a few short weeks before his assassination and the topic was not race but classism and Vietnam. With his words, this great American inspired thousands to join the peace movement, to bring our troops home from a war which the government had decided was hopeless.

We are now faced with the fact that tomorrow is today. We are confronted with the fierce urgency of now. In this unfolding conundrum of life and history there is such a thing as being too late. Procrastination is still the thief of time.

Life often leaves us standing bare, naked and dejected with a lost opportunity. The "tide in the affairs of men" does not remain at the flood; it ebbs.

We may cry out desperately for time to pause in her passage, but time is deaf to every plea and rushes on. Over the bleached bones and jumbled residue of numerous civilizations are written the pathetic words: "Too late." There is an invisible book of life that faithfully records our vigilance or our neglect.

I took Dr. King's words to heart and knew that it was time to fight the good fight. Living in America no longer was about going along to get along. If I really wanted to fix matters and protect my home from the ravages of climate change and voter apathy, the time to act had arrived and I should let no obstacle hamper me.

Stand For Something Or Fall For Anything

It is a unique sensation when an introverted person contemplates running for office. The best word I can use to describe it is humbling. Knowing that one's views are correct and abide with the wishes of the framers of our Constitution is one thing but it is quite another to determine how to share those views with strangers in a compelling way so that they will donate money, or volunteer their time or, most importantly, make the effort to go to their local polling place on election day and draw an arrow next to that person's name. Ponderous souls such as I am easily fall victim to the rectitude of our inner voice. If a notion sounds good to us, everyone else will feel likewise.

Understanding that I do not have all of the answers is the beginning of wisdom, particularly political sagacity. It is for this reason that I hope to have with you, dear reader, not just a chat but a higher conversation, a dialog in which I learn your beliefs, so that I can represent them adequately, should I be elected to Congress, and in which I convince you of the basis of and reason behind my convictions, especially those which are derived from the absolutes given to us by the Founding Fathers. Let us begin, then, with the core, those most pressing issues of this time in America's history, which have the greatest likelihood of shaping the conditions under which the generation which succeeds us will live and prosper.

Global Climate Change

Humanity is the main cause of climate change. Global warming is the primary agent of climate change and the earth is warming because of three unsustainable activities:

1) The burning of fossil fuels for energy, including coal, natural gas and petroleum;
2) The destruction of forests, both through logging and overdevelopment/urban sprawl; and
3) The poisoning of our oceans through chemical runoff from cities and farms and ecological imbalance caused by destructive fishing practices

In the same way that the checks and balances of America's system of government act as a pair of lungs for the society, the planet has a pair of lungs, its forests and its oceans. With each passing day in which we fail to reform our agricultural, aquatic and industrial practices, the earth smokes the equivalent of two packs of unfiltered cigarettes. In other words, the planet already has bronchitis and soon, it will have emphysema

Poverty

As a Christian, I know that I should refrain from judging the sins of others. However, as a patriot who is fed up with seeing his countrymen suffer, I struggle with a different challenge: the greater misdeed done in the name of religion— the reckless and unsustainable kickbacks to wealthy campaign donors in the form of tax reductions for top income brackets or the wholesale disregard for the suffering of millions who toil with inadequate housing and access to affordable medical care.

> *He that oppresseth the poor to increase his riches, and he that giveth to the rich, shall surely come to want.*

> — *Proverbs 22:16*

Let me be clear, however, that I do not struggle with the knowledge of how America should conduct herself in the care for her citizens, especially the most vulnerable. Universal healthcare is our birthright, every bit as much as is a high school education and when we fail to live up to that birthright, we suffer as a culture, as an economy, as a military power and as a society. According to the *American Journal of Preventive Medicine*, people who live in deep poverty, earning less than half the annual income which the federal government describes as the poverty line, has been the fastest growing segment of our society since the year 2000. Long-term hand-outs to able individuals are wrong but permitting such destitution not only to exist but to swell is far worse. It is inhumane and uncivilized.

I recognize that few if any other issues confronting America at this time in her history generate more staunchly drawn ideological lines. As you can gather from the previous paragraph, I have a very specific perspective. What those who disagree with me must realize, though, is that our healthcare crisis no longer affects just the poor and has ceased to be confined to finance. The lack of universal healthcare has swollen the ranks of the uninsured. Those people have turned to hospital emergency rooms for care in expanding numbers either because outpatient urgent care is unavailable where they live or they cannot afford to pay for such care and know that hospitals cannot turn them away if their condition is life threatening.

Because more and more patients cannot pay some or all of their bill, hospitals are forced to absorb more and more expenses. Some are better able to do this than others. The unfortunate result is that emergency rooms have begun to close. In the case of Central Florida, I have seen the region's only Level 1 trauma center nearly shuttered twice since I moved here in 1991 because of funding issues.

More emergency room patients + fewer emergency rooms = recipe for disaster

Emergency rooms are not allowed to inquire about a person's health insurance coverage during triage and rightly so. Priority 1 must be assessing the patient and administering the correct care as soon as possible, based on the patient's medical condition. If Jane Doe is involved in an automobile collision which destroys her car, she may not have insurance information on her person. Should doctors wait to determine her medical needs until she can confirm what her deductible is? Certainly not!

While the only civilized choice, this standard of care first, charge later has a large downside; it relies on the honor of patients to pay for their care after they receive it. People know that if they are ill, they can visit the emergency room and be seen eventually. Over time, this knowledge has created very long wait times at hospitals. Such delays do not discriminate between insured and uninsured patients. There is simply more demand for beds than there is supply. Wealthy people with no deductible on their health insurance often are stuck waiting to see an emergency physician for just as long as the homeless. Providing basic care has become a matter of basic decency.

The only solution is universal healthcare so that hospitals can base their rates on actual expenses plus a small profit margin and be assured that they will be paid. A steady revenue stream which exceeds expenses is the foundation of most any viable business. It allows for projects and growth plans. If the United States is to remain a premier destination for business and tourism, travelers have to know that competent healthcare is available if they should need it while in greater Orlando and workers who consider moving here in order to seek employment in our vast service industry must know that they can find affordable healthcare for their families.

As the matter of healthcare in the United States stands today, we spend vastly larger sums per capita than other industrial nations yet have poorer health when compared in aggregate. What's more, whereas America in general and Florida in particular once were attractive to foreign doctors who wished to practice outside their homeland, we now have growing shortages of good doctors. Robyn Shelton of the *Orlando Sentinel* summarized the matter very well in the newspaper's edition of June 15, 2006: "The nation's emergency-medical system is in crisis, with crowded ERs turning away ambulances, patients waiting hours to be treated and a shortage of on-call specialists."

Dr. Michael Marmot of University College London is studying the question of better overall health in the United Kingdom for a smaller public investment. One month before Robyn Shelton's article ran in the *Orlando Sentinel*, Dr. Shelton posed an interrogatory condemnation: *"Why isn't the richest country in the world the healthiest country in the world? It's not just how we treat people when they get ill, but why they get ill in the first place."*

Poverty simply cannot be seen as an issue which poor people who can't or don't want to work must battle. It affects everyone and while it was easy to ignore for many years, it has grown to impact healthcare for every American who uses or expects to be seen by public emergency rooms. I fear that comprehensive reform will not take place until hundreds of people die needlessly each year from a crumbling safety net and I promise to fight for that reform in Congress for as long as it takes. It's the only just option. I dare to care.

Runaway Executive Compensation

At the same time as we struggle with abject poverty within our borders, we must consider policies aimed at runaway compensation packages for corporate executives. In 2005, the top executives of many public companies earned in a day what the average employee of that same company earned in a year, a ratio of about 300:1. Capitalism is part of what made America great but avarice is not a core national goal. Furthermore, when companies decry attempts by government bodies to require that they offer affordable healthcare to all of their workers, their argument that they cannot afford to do so must be tempered with the ratio of average pay to executive pay.

If you woke up this morning with more health than illness, you are more blessed than the million who will not survive the week.

If you have food in your refrigerator, clothes on your back, a roof over your head and a place to sleep, you are richer than 75% of this world.

If you have money in the bank or in your wallet, you are among the top 8% of the world's wealthy.

If you hold up your head with a smile on your face and are truly thankful, you are blessed because the majority can, but most do not.

— *Author Unknown*

Central Florida consistently ranks as the top tourist destination in the world. We export hope and optimism. How do we help the rest of America do the same thing? We need a resonant voice in Congress to share our story with the nation and the world. I wish to be that voice but the story I tell is tarnished when so many residents of greater Orlando live in poverty.

"More than 14,000 people in Central Florida—the equivalent of Winter Garden—must decide every week between paying the rent or eating."

—Orlando Sentinel, 6/6/2006

"If you're not getting enough to eat, eventually it hurts your ability to work and stay healthy. It's a downward spiral."

—*Dave Krepcho, Second Harvest Food Bank of Central Florida*

"At eight bucks an hour, there aren't enough hours in the week to work and live well."

—*Creighton Knight, Greater Orlando Food Bank*

Reproductive Freedom

How many more women must be elected to Congress before we treat them as equals? How many more women must become executives of large corporations before we close the gender gap? If women have the right to vote and can wear the uniform of the United States as a part of our military, then surely they deserve the same control over their body which men enjoy. What's more, no definition of "Life, liberty and the pursuit of happiness" would be complete without the freedom to choose if and when to bear a child.

Are there times when a women's obstetric medical decisions should be restricted? Yes, there are small numbers of limited cases in which it is clear that a woman behaves irresponsibly but we must not use those rare instances as justification to shatter the essential right to confidentiality between a patient and her physician.

You may wonder how I would vote on attempts to restrict reproductive choice if elected to Congress so allow me to be clear. Reasonable people can draw reasonable conclusions from an abundance of facts but the debate on a woman's right to choose has been overly narrow for too long. I will not vote to authorize new limits on what is first and foremost a medical decision if the bill which proposes them does not provide a sweeping overhaul of sexual education in this country. If we educate our young adults about all of the facts regarding the human reproductive process, such as the teaching I received in high school from my 10th grade science teacher, we will reduce the overall demand for post conceptive procedures.

Love Decriminalized

The decision to marry often can be made through rational thought. It can be a choice. What seldom is a choice, however, is whom we love. The human heart is capricious in deciding to give itself to another and government has no business trying to interject its standards into matters of the heart. We can restrain people who lack proper judgment due to age (children, specifically) or mental incapacity but on the whole, consenting adults should be free to choose their mate. Sadly, the American standard for matrimonial bonds is a patchwork of inconsistencies. The most controversial of these inconsistencies in current events is that of civil unions and marriage for homosexual couples. My position on the matter is simple.

Sexual orientation is genetic. My heterosexuality was set at the same time the color of my hair was set, at conception, when I was but a zygote. I could dye my locks blond but genetically, I still would have dark brown hair. Denying consenting adults the right to form a legally-recognized bond because their genes made them homosexual is flat wrong and contrary to associational freedom as set forth in the Bill Of Rights. Hence, I support civil unions in all American territories. Civil unions should be enforced at the federal level and all states should be required to recognize civil unions from other states. Lastly, gay couples who enter into civil unions should enjoy the same rights and responsibilities as straight couples, including inheritance and hospitalization rights.

I oppose, however, compelling houses of worship to conduct weddings for gay couples. While I am unchangeably convinced that sexual orientation is genetic, I also am convinced that the strength of our democracy is inversely proportional to the distance between church and state. That is, our collective future shines brighter and brighter as government involves itself less and less in organized religion. Consequently, even though my particular faith, the Episcopal church, continues to wrestle with the issue of weddings for gay couples, I know that the courthouse which is located just two city blocks from my church should perform civil ceremonies for homosexuals.

Gay marriage also contains an element of states' rights. Ever since the founding of our great nation, matrimonial law has remained within the purview of individual states. It is the states which decide at what age a child becomes an adult and may marry. It is the states which decide whether a suit for divorce requires evidence of abuse or neglect. Sadly, for decades after the end of the Civil War, states were allowed to keep dreadful prohibitions against interracial marriage, miscegenation, on the books. I disagree with those states which have altered their constitution to outlaw gay marriage but I would not vote to overrule the decision of the voters in that state. At the same time, however, as a Congressman, I would vote against any bill which proposes to amend the Constitution of the United States banning gay marriage.

The same spirit of restraint, the spirit of minimizing federal interference in the affairs of its citizens means that civil unions must be available from sea to shining sea but that it is up to the states to decide when and if they will grant or recognize gay marriage ceremonies.

Restraint and decorum also dictate that hypocrisy has no place in the debate over universal civil rights. Elected officials who campaigned on the fact that they are Christians yet were unable to remain married to their spouse relinquish their right to decry the marital wishes of others. Scott Maxwell of the *Orlando Sentinel* put it succinctly in his web log on July 19, 2006, *"There is something downright strange about hearing proclamations of the sanctity of marriage from so many politicians who've left their wives."*

A Politicized Pentagon

The Congress has three roles with regard to the deployment and operation of the military of the United States. First, it is coequal with the executive in declaring war. Second, it allocates funding for military operations. Third, it oversees the general operation of our armed forces to see that the leaders are fulfilling their commitments and that the men and women who wear the uniform of the United States are treated justly at all times.

A number of members of the United States Congress who voted in favor of the 2002 authorization for hostilities against Iraq have expressed regret at doing so. The reasons given for those regrets generally fall into just two categories. The first is that the intelligence reports were inaccurate or exaggerated to support claims that the Hussein regime posed an imminent threat to American interests and those of its allies. The second is that George W. Bush committed to exercise restraint but invaded hastily thereafter.

The peace marches which took place early in 2003, during

the period known as the run-up to the war in Iraq, impacted me deeply and I knew that our planned invasion was wrong on many levels. However, as you read this, many years have passed since then and you have only my good word to know how I would react as a Congressman were I faced with a similar floor vote. As a method of knowing my heart, of knowing the decision process I will follow if the 44th President of the United States seeks to declare war, I offer the following formula, informed by the mistakes of 2002 but also by an intense focus on the havoc which war wreaks on American troops and their families.

The history of my family since the 1940's can be summarized rather easily: most of my relatives became either preachers, teacher, or soldiers. Both of my grandfathers were men of the cloth. Both of my parents were teachers. My maternal uncle Howard served in the United States Army for many years, including in Vietnam. In fact, one of my most distinct childhood memories is of being in the basement of our family home in Pennsylvania. We had gone there to escape the summer heat and my mother telephoned Uncle Howard while he and his wife Joan were stationed in West Germany. I was but 9 or 10 years of age at the time and could not understand where West Germany was or why my relatives would live there. As time passed and I came to recognize the effects which long overseas postings have on families, I swelled with pride at my uncle's dedication.

> *"History teaches that war begins when governments believe the price of aggression is cheap."*
>
> *—Ronald Reagan*

When I think, then, of striking a balance between authorizing war as a Congressman and safeguarding the lives of the men and women who serve in our military, my process will be to weigh the verisimilitude of the executive branch's case against the inevitable costs of blood and treasure. The very first measure I will take is whether the President is a combat veteran. I then will measure the military record of the Vice President and Secretary of Defense. Lastly, I will scrutinize the plans for victory, peace and withdrawal. I need not have attended West Point or the Army War College to know whether a strategy makes sense on its face and whether, in good conscience, I can ask the residents of Florida's eighth congressional district to support it.

"The military doesn't start wars. The politicians start wars."

—General William Westmoreland

A politician's record in supporting our troops is fair game during campaigns. However, criticism of a Congressman's voting record on military affairs should be limited to the three responsibilities which I set forth earlier: authorization of war, allocation of funding and oversight of general operations. One of the most vital aspects of those operations is the treatment of our dead and wounded. For too long, funding for veterans affairs has been used as a political pawn and it must stop. Such funding should be guaranteed year to year.

In the era when America drafted most of its sailors and soldiers, one of the benefits touted by recruiters and retention offers was universal healthcare. They reminded draftees that fighting for Uncle Sam meant that he always would take care of those in uniform. No longer! We must return to those days and stop excluding from Pentagon budgets the care of our veterans when they need us most. To do anything less is a moral failure.

The largest segment of America's homeless population is comprised of veterans of our military. When I learned that statistic, I was nauseated. Since then, the problem has worsened because they have been shortchanged on account of runaway expenses in Afghanistan and Iraq.

> *"Almost half of America's 2.7 million disabled veterans receive $337 or less a month in benefits, according to the government. Fewer than one-tenth are rated 100 percent disabled, meaning they get $2,393 a month, tax free."*
>
> — *Associated Press 7/5/2006*

Public Education

It was the late, great Thomas Jefferson who gave us the principle of a basic education for all. Sadly, his vision began to lose steam after World War II. The fight against Communist infiltration became nearly all-consuming. Its ripples were felt in the public curriculum. To add insult to injury, classroom teaching methods and technology failed to keep pace with cultural changes. By the 1990's and early into the new millennium, the MTV generation had grown up accustomed to almost constant sights and sounds. Auditory and visual stimuli are difficult to replace but in many cases, America's public schools did not even try.

To be fair, the United States public education system had been poor for a full decade when No Child Left Behind cleared Congress and George W. Bush signed it into law. The main causes of its failure are well known. It exacted high standards but provided inadequate funding. If Newt Gingrich had remained Speaker of the House, he might well have referred to the new requirements as unfunded mandates. Worse, though, is the fact that No Child Left Behind converts teachers into bureaucrats. They are forced to spend so much of their time preparing their students for standardized tests that little time remains for learning.

The solution, not surprisingly, begins with more money. However, that money must be used judiciously. Teacher salaries are the right place to begin, with strict health and safety standards for school buildings following in close formation. Most other industrial nations recognize that certain jobs, which tend to affect the fabric of society, are special and should be treated as such. Most states pay their judges well enough that they can afford to raise a family on a single income. Sadly, very few do the same with teachers.

In the same way we would not expect nurses to care for people in life and death situations while earning little more than minimum wage, we should not expect the people with the second greatest impact on the prospects for a bright future of our children to scrape by, earning barely enough to keep a roof over their head and food on the table. Florida already suffers from a shortage of teachers in many areas and hiring them near the bottom of the economic totem pole does little to remedy the situation.

Money for salaries and adequate school buildings only is the beginning, however. The real essence of the problem is inspiration. I was an honors student on and off in high school but never considered myself a top student. I was fortunate to have left public education with a decent head on my shoulders but I also found class dull and boring most of the time. There were rare occasions when I found that the material regarded my interests or had relevance with my adult goals. During those times, I sat in my chair with a heightened sense of awareness, processing information faster, retaining a greater percentage of the lesson and feeling exhilarated. It was during those moments that I felt inspired.

Why do teachers, people who obviously have more than enough intelligence to enter the business community and earn top salaries, choose to teach? It is because they are energized by the proposition of shaping the next generation of Americans. It is because they know how vital to their success it was to have a decent education and they want to assure that others have that benefit. It is because they know that children who are taught well grow up to become doctors and lawyers and (gasp!) politicians. It is because the idea of teaching inspires them.

Inspiration is the key. Accountability is important and practices such as social graduation leave homeowners wondering how their tax dollars are being spent but no program for education reform will succeed until teachers are free to teach, until schools have 21st-Century equipment and methods and until students are made to realize that staying in school is fun and rewarding.

Iraq

George W. Bush's invasion of Iraq was illegal and immoral. I cannot state the matter more concisely than with that single-sentence declaration. Bush's war is illegal because he lied to Congress in order to obtain their authorization to use force, because Cheney strong-armed the Central Intelligence Agency and its counterparts to contort intelligence reports. In British parlance, the term is "sex up." The war is immoral because it is illegal, because its post-combat strategies were amateurish, because every criticism of the campaign both at home and in allied nations was met with political maneuvering and because

Bush's people have allowed their cavalier incompetence to create a new Vietnam which the world will need decades to heal.

Mr. Bush's 2000 and 2004 campaigns included targeted messages for people of faith, particularly evangelical Christians. They are my brethren but their righteous claim on protecting human life conveniently overlooks the apocalyptic numbers of casualties who otherwise would be alive today had we not invaded, most of them non-combatants, many of them women, children and infants. Who will atone for their deaths? Who will communicate with the hundreds of thousands of refugees who have fled their ancestral lands to neighboring countries to explain that they were forced to leave as part of a political strategy which became Trail Of Tears II?

America must come to terms with the human costs of Iraq. The beginning of that tally, of course, is an accurate accounting of our losses but followed very closely by the number of people who have died, been wounded or lost their home because we capriciously destabilized an entire region of the earth. What would I do as a United States Congressman? I would do three things:

1) I would start the process of creating a fund and administrative infrastructure for the payment of war reparations, conducted through the Department of State;

2) I would compel the Pentagon to disclose to the world the true location and scope of permanent military posts which we have erected in Iraq, which are supposed to serve as our bases of operation after hostilities cease; and

3) I would recommend to the legitimate Iraqi government

that it consider nullifying all petroleum concessions and contracts granted to large corporations during the period when the United States controlled the nation. If the Iraqi people wish to strike new arrangements with those companies, that is their own internal affair but the devastation which our combat activities have rained down on the Iraqi people should not be compounded by Enron-style oil deals which we granted when no civilian groups had a say in the matter.

Habeus Corpus

Why did the American colonies declare their independence from the British crown and revolt? Was taxation without representation the main cause? Was it the forceful suppression of dissenting views? Perhaps it was the policy of stationing military garrisons in civilian housing, usurping the rights of the homeowners to have dominion over their own dwelling place? In reality, the list of grievances against the King of England is lengthy and compelling and the Declaration of Independence makes the case very well that the colonists had tolerated enough, that no civilized people should be forced to brook such malfeasance.

Despite the grave misdeeds committed against their countrymen, most American colonists were content to live under a system of government which bore a strong resemblance to British common law. Indeed, many of the colonial constitutions remained largely intact when they became state constitutions after the War of Independence and the national Constitution borrowed heavily from many of those state constitutions. Why is that? The Anglo-Saxon system of government guided by the will of the people was nearly unique in Western Europe. The framers of the Constitution knew that the British system itself was well conceived and the cause of the discontent of our colonists was the despotic behavior of the king.

One of the most fundamental precepts of the British system is the right of Habeus Corpus, the right of a prisoner to meet his/her accuser and know what charges have been brought. Indeed, Habeus Corpus is so fundamental that its existence and application predate Magna Carta, considered to be the basis for the success of the English Constitutional Monarchy because its final version clearly defines the boundaries of the national legislature and the power of the Sovereign, the royal head of state.

Habeus Corpus was part of the intent of our Founding Fathers and, shortly after the ratification of our Constitution, it was enshrined in the Fourth Amendment. Sadly, though, it was abused and curtailed during the first two centuries of our nation's history. The Supreme Court reaffirmed and clarified its use in the Miranda versus Arizona decision, giving us the phrase, "You have the right to remain silent." and the right to a public defender if you cannot afford to retain legal counsel. That is, until George W. Bush decided that he is the decider, that he decides whether you are entitled to these fundamental rights under the laws of a just society.

The Military Commissions Act of 2006 was an abomination against the sacrifices of every American who fought and died to free us from British tyranny. Alas, it was not the first time Habeus Corpus has been suspended. John Adams was wrong to ignore Habeus Corpus when he demanded the powers of the Alien and Sedition Acts. Abraham Lincoln was wrong to suspend Habeus Corpus during the Civil War and Franklin

Roosevelt, one of the greatest Democrats ever, creator of the New Deal, was wrong to use it against Japanese-Americans. History has taught us, albeit not in the immediately succeeding years, that dispensing with Habeus Corpus is wrong, each and every time we have done so. However, when we examine George W. Bush's myriad errors during his time in the White House, it is clear that he was absent for every one of his high school history classes and thus, we should not be surprised that he requested and signed into law the latest assault on basic human decency, the Military Commissions Act of 2006.

As a United States Congressman, I will fight to restore Fourth Amendment rights for everyone, citizen, alien and visitor. We cannot combat despotism by embracing a behavior which we decry whenever another acts in such a manner.

Torture

The repeated and feeble rebuttals of George W. Bush that the United States neither condones nor uses torture are, at best, an opaque tissue of lies and, at worst, malfeasance. Under his reign, unspeakable horrors have been committed in our name, sometimes by Americans only following orders but other times by shady operatives of nations which we call allies but which have no such legal precedent as the Fourth Amendment. The Bush administration has made it clear that reliance on the executive branch to adhere to the Constitution without constant vigilance from the House of Representatives and Senate is naïve. Hence, as a member of Congress, I would fight to codify the limits of interrogation which may be used by the United States government as well as by any other person or agency which our government may employ or utilize to interrogate our prisoners.

I recognize that protecting America's national security sometimes necessitates arresting and holding suspects indefinitely, especially if they pose a flight risk. However, there can be no justification for torturing anyone, under any circumstances and the time has come to codify a penal system for punishing those who would conduct torture, whether under orders or of their own volition.

American must not torture!

Global Warming

Let's return to my top issue, the greatest threat to our nation and our world.

Part of the reason Vice President Al Gore's film version of "An Inconvenient Truth" became the third-highest-grossing documentary of all time, won an Oscar and became a vital teaching tool in the United Kingdom is that it disturbs while it educates. It limits the boundaries of possible outcomes of inaction to the scientifically inevitable. There are predictions from very wise climatologists that the year 2100 will be the end of human civilization as we know it and, of course, predictions that all will be well, that humanity is free to drive sports utility vehicles and suck the planet's petroleum reserves dry with abandon.

Mr. Gore is correct, of course, in his explanations of the causes, effects and remedies. Political will is the solution and it is a renewable resource. However, what I have not observed is a promulgation on the part of environmental groups of a strategy for assuring that such political will is tapped and sustained. We must preach such a strategy loudly and often because every single world citizen worried about climate change must remember that the Carlyle Group, Exxon, Halliburton and their friends will throw hundreds of millions if not billons of dollars of influence at anyone who dares disassemble the oily empire which they have built.

How, then, do we go about applying political force in Washington to address global warming? Even if we find thousands of people who demand that their leaders take decisive action and immediate, sweeping changes are brought to bear, no lasting solution will exist without leadership from the Congress of the United States. As you will read in the final pages of this treatise, I have a specific formula for how we go about electing people to Congress who will act in accordance with the wishes of their constituents most of the time. However, solving the problem of global climate change will require more than that. It will require the widespread use of clean technologies. The good news is that those technologies exist today and have a proven track record.

Our ultimate goal should be similar to the energy revolution now underway in the North Atlantic nation of Iceland, a move toward a 100% hydrogen economy. One of the rare instances in which George W. Bush has taken a correct and pro-environment position was the 2003 State of the Union Address in which he committed the United States to hydrogen. The funding which resulted from that bold announcement has been grossly inadequate yet real progress has been made and he deserves credit for it.

What should have happened, however, was the redirection of the billions of dollars of pork given to Exxon and the rest of the oil industry in the Energy Policy Act of 2005. Between grants and loan guarantees, Exxon alone received several billion dollars. That donation of tax money to a public corporation actually should have gone into incentives to build a hydrogen distribution network along Interstate 10 and new bio refineries to convert cooking waste into BioDiesel. Happily, it is not too late to do so and, in small pilot programs all over North America, entrepreneurs are doing just that.

Hydrogen as a transportation fuel suffers in a scenario of the chicken and the egg. There are very few vehicles on the road capable of burning hydrogen because there are even fewer hydrogen fueling stations. At the same time, there are very few hydrogen fueling stations because so few cars are on the road today in need of a fill-up of H2. This difficult situation is exactly why a government program is the ideal way to begin. When no company or person is willing or able to tackle a large national problem, it is appropriate for the government to allocate public funds to solve the problem. It happened with the Interstate Highway system. It happened with car safety belts. It happened with the Internet.

Even with proper funding, though, we will need at least a generation in order to eliminate petroleum as a transportation fuel. Fortunately, BioDiesel can play a vital role as an interim technology. Whereas ethanol (called gasohol the first time America toyed with it) presents a problem in balancing the needs of grain for food versus fuel, BioDiesel is simple. Using it in vehicles requires little or no modification; the fuel may be blended with petroleum fuel in any ratio and it can be produced from used kitchen material such as cooking oil, which otherwise must enter our nation's waste stream.

The other compelling reason to fund national adoption of BioDiesel as the fuel of choice is that our economy is much more vulnerable to shortages of diesel fuel than it is to shortages of gasoline. If the country ran out of gasoline tomorrow but still had diesel fuel, it could continue to function. Consumers would be very unhappy, of course, but the country could remain open for business. The reverse is not true. The overwhelming majority of surface freight systems are powered by diesel fuel:

buses, ships, trains and trucks. If we suddenly had no more diesel fuel, those vehicles could not be supplanted quickly or easily. If those surface freight systems ran on BioDiesel, however, America's domestic commerce would have a self-sufficient fuel supply.

Consequently, as a Congressman, I would advocate the following steps to upgrade our energy supply and combat global warming:

1) By the year 2020, eliminate petroleum diesel fuel from the American energy system and encourage the other G8 nations to do likewise, replacing it with 100% BioDiesel;

2) Ban the construction of new electrical generation plants which are powered by fossil fuels or nuclear fission, permitting only biogas, solar and wind-based production; and

3) Establish a nationwide renewable energy portfolio which requires that all electricity sold be comprised of at least 20% green power, either from biogas, small hydroelectric, solar or wind-based production by the year 2020.

The knowledge and hardware needed to make such sweeping reforms possible and affordable exist today. The only absent component is political will. Money always favors the status quo. Consequently, success will be based on the power of our convictions and the number of people who share them.

Political Action Committees

Perhaps the most important precept of the Bill Of Rights is the concept of the freedom of speech: that personal liberty can take many forms and the one which causes great consternation to some is the importance of protecting the right of others to an opinion even if one disagrees with that viewpoint. In order for the national government to function properly, however, we must allow the First Amendment to blossom to its fullest flower. The temptation to curtail free speech must be eschewed. It must be resisted with vigor.

Political Action Committees can help to focus the din of millions of constituents into a palatable choir. Few if any Americans could say with a straight face that no lobbyist or lobbying firm espouses an offensive position. Indeed, it is the blend of dissenting views which gives us the confidence that our collective decisions are the best for all. What happens, though, when that choir becomes dissonant or when there are more choristers than there are directors, such as the situation which exists today? Harmony becomes a strident scream.

Several attempts have been made in recent years to silence the voice of lobbyists in the nation's capital. For my part, I certainly share the conviction that many lobbyists are the mouthpiece of deep-pocketed contractors waiting to milk taxpayers for lucrative allocations. However, they are not all bad. Indeed, I learned through my experiences with the Sierra Club that a small but not insignificant number of them are good people helping to assure that the business of all Americans is done.

Where, then, is the balance? If the work of lobbyists is protected by the Constitution, how do we diminish their efforts? I have an answer. It may an incomplete solution but

I know that it is the right place to start. Politicians must refrain from accepting gifts from or being unduly influenced by lobbyists. Politicians must keep lobbyists at arm's length. Hence, I commit here and now to do the following if elected to Congress:

1) If a lobbyist wishes to meet with me in an official capacity, such a gathering must take place in my Capitol Hill office or my district office. Furthermore, meetings with lobbyists must be open to constituents and any American taxpayer.

2) I will conduct no official meetings with lobbyists outside the confines of item 1 above.

3) I will receive no remuneration or anything of value from lobbyists. If I choose to appear at a public event sponsored by a lobbying firm, I will do so at my own expense unless directed to do otherwise by government officials.

4) Lobbyists will be unwelcome at any official gathering during which I draft legislation. A lobbying firm will hold the same sway over how I write bills as an individual taxpayer does.

Political Action Committees are a purer form of lobbying firm as I view the matter. While some of them serve as slush funds for large corporate donors attempting to short circuit the design of our democracy, many of them simply help to coordinate the wishes of blocks of voters so that legislation can be affected positively. For my part, I have donated to many Political Action Committees and plan to continue doing so. Such plans, however, do not mean that I have to accept campaign donations from them. In short, I pledge to accept no money of any kind from Political Action Committees, not one dime.

Such a pledge begs a question, however: if Political Action Committees should not contribute to individual campaigns, what should their role be? Lobbying Congress and convincing voters who agree with their position on an issue or an array of issues is not an inexpensive proposition. I have no objection to their use of the airwaves and the halls of Capitol Hill to propagate their position, provided that it is done with full disclosure and that the elected officials with whom they interact do not become cozy with those interests. What we have today, however, is far, far worse than the delicate balance I envision. It borders on prostitution and it must end. I am confident that my time in office would demonstrate that it is possible to protect freedom of speech for all while simultaneously protecting the rights of taxpayers not to have their national legislature corrupted by snake oil salesmen.

$100 Per Month

No contributor may give more than $100 per month to my campaign.

The United States House of Representatives is supposed to be the people's house. Its procedures and its 2-year election cycle were designed by the Founding Fathers with great deliberation in order to give every American an equal voice in national debates. An equal voice means irrespective of social station or stature—irrespective of wealth. Limits on contributions to federal campaigns of late have helped force candidates to engage voters more directly but a potential donor's ability to write a check for the full campaign limit in a lump sum still permits an imbalance. I will impose a personal limit to narrow the influence of wealthy contributors.

When the campaign finance reform law known as McCain-Feingold was passed, one of its goals was to assure that individual contributors are not ignored and that well-funded interests cannot sway a particular election. A method for doing so as set forth in that new law was the concept of a small contribution. A round number of $100 became the benchmark. I realize that running for federal office will be quite expensive but I want to be sure that no one has unfair access to me as a candidate or Congressman.

Florida's eighth congressional district is a study in contrasts. It is drawn in such a way that many of the region's wealthiest neighborhoods are contained fully within its borders but also just outside the poorest neighborhoods. Many of those lower-income areas are literally right across the street from areas which are in the district. A voter who lives in a lower-income area might find that the candidates for his or her district would not be overjoyed with a contribution of $100 or less but would receive the same smile, handshake and thank you if he or she gave the money to my campaign as a donor from Bay Hill, College Park or Windermere would receive because it's all the same to me.

Setting a limit of $100 per month from any donor assures the grassroots nature of my campaign, especially its funding, and makes it that much more difficult for my opponents to claim that I represent special interests. Grass is green. Chlorophyll, the substance in plants which allows them to produce oxygen, is green. Money is green. A grassroots campaign about protecting the natural environment which only accepts small financial contributions is about as green as a campaign can be. It is no coincidence, then, that green is one of my campaign colors.

SECTION 1
Conclusion

It began as heartache and anxiety for the future and became a mission, a mission to change how elections are run and how Capitol Hill is run. The Founding Fathers created a path for how America should conduct itself and how the national government should protect the people yet heed their wishes. John W. Dean, former Nixon White House counsel, recently said that, "Today's Republican policies are antithetical to bedrock fundamentals." The only way to put our nation back on track is to focus on the future, on an inclusive form of self-government, such as the revolutionaries of more than two centuries ago espoused when they took up arms against Great Britain. I view that track as not unlike a pathway, the Progressive Pathway, not always straight, not always smooth, but always a joy to behold.

Join me.

www.ingramcontent.com/pod-product-compliance
Lightning Source LLC
Chambersburg PA
CBHW020347290526
45785CB00005B/2184